KV-192-320

SPUR PUBLICATIONS

POULTRY FANCIERS' LIBRARY

Poultry Colour Guide

DEDICATED TO THE POULTRY CLUB AND ITS MEMBERS
WITHOUT WHOSE EFFORTS OVER THE PAST 100 YEARS
THE MANY BEAUTIFUL BREEDS WOULD NOT BE IN
EXISTENCE . . .

IN THIS CENTENARY YEAR 1977

FIRST EDITION 1977
SECOND EDITION 1979
THIRD EDITION 1985

POULTRY COLOUR GUIDE

Covering Large Fowl, Natural
Bantams, Ducks, Geese, Turkeys
and Guinea Fowl

DR. J. BATTY

Past President
Old English Game Club

With illustrations based on the portraits by
CHARLES FRANCIS

Distributor:
NIMROD BOOK SERVICES
P.O. Box 1
Liss, Hants, GU33 7PR
England

© J. BATTY   1977, 1979, and 1985

ISBN 0 947647-03-1

This book is copyright and may not be reproduced in whole *or in part* (except for review) without the express permission of the publishers in writing.

Printed in Hungary
Publisher:
NIMROD BOOK SERVICES
(Fanciers Supplies Ltd)
P.O. Box 1
Liss, Hants, GU33 7PR

EDITORS' FOREWORD

Shortly after Spur Publications was established we asked Poultry Fanciers to complete a questionnaire indicating what kind of books they would like. The majority stipulated that coloured illustrations were of prime importance and, therefore, we proceeded to introduce these into a number of books which were published.

In attempting to cover the major breeds and varieties in colour in the present book there were many difficulties. In some cases, there were no good specimens available on which to model type and colour. Advice of fanciers had to be sought and references made to old books and journals as well as *British Poultry Standards*.

We were fortunate in being able to co-operate with Charles Francis in placing on record the breeds and varieties of poultry—large fowl, bantams, ducks, geese and guinea fowl. His task required skill and patience in painting the birds under our guidance.

We believe this is the first attempt to illustrate the full range of domesticated poultry in their full beauty. Indeed, not since Lewis Wright's *Illustrated Book of Poultry* and Harrison Weir's *Our Poultry* has a major poultry book been written, which includes coloured portraits.

Poultry-keeping is a fascinating hobby. Yet many people are not even aware of its possibilities. Fewer still even appreciate the multitude of colours which are available. This book is intended for those who wish to know more of these practical, beautiful birds. It is a guide for the poultry fancier who has for so long required a work of reference.

M. BATTY   J. BATTY

FOREWORD TO THIRD EDITION

A further edition shows the continuing interest in poultry breeding. Moreover, the movement towards the revival of free range poultry, and the criticism of battery cages, appears to suggest that more robust birds, suitable for outdoors, will be essential. This could lead to the keeping of standard breeds for commercial purposes which used to occur many years ago.

LISS
1985   J. Batty

ACKNOWLEDGEMENTS

We would like to thank all fanciers who assisted in some way in making this publication possible. This wish also extends to writers of books and contributors to journals, past and present, particularly *Fur and Feather,* and *Poultry World.*

Reference was made to *British Poultry Standards* and readers who wish to acquaint themselves with the full descriptions of the breeds and varieties are referred to that source. We also consulted the coloured plates issued by the "Feathered World" and "Poultry", sadly both now part of the history of the Poultry Fancy.

Visits were made to see birds kept by fanciers. Those wishing to acquire stock are advised to refer to the *Poultry Club Year Book* which contains names and addresses of breeders and Secretaries of the various breed clubs. Special mention should be made of the work being done by Mr. Alistair McAlpine of Hartley Wintney, Hampshire, who has established a fine collection of the rarer breeds of poultry including water fowl. He is doing much to make stock available to the fancier who wishes to make a start with a particular breed.

Finally, but by no means least, we acknowledge the invaluable assistance given by four dedicated poultry fanciers:

Miss Veronica Mayhew, Member of the Council of the Poultry Club and Hon. Secretary of the Belgian Bantam Club.

Mr. Charles Mayhew.

Mr. Peter L. Parris, Member of the Council of the Poultry Club and Hon. Secretary of the Welsummer Club.

Dr. W. C. Carefoot, Member of the Council of the Poultry Club and Hon. Secretary, Partridge Wyandotte Club (see page 12).

After the paintings of the birds were completed the first three named spent time in studying the colours and recommended where corrections were necessary. If a high standard of perfection has been achieved then they contributed in a significant way. However, they are not responsible for any errors or omissions which may exist. The author and artist must carry that burden.

CONTENTS AND INDEX TO BREEDS

LARGE FOWL

Breeds which exist in Bantam form are indicated by an *

THE POULTRY EVOLUTION
BIRDS FROM AFAR*

Domestic fowls have existed for centuries. They have evolved from the Jungle Fowl, but precisely how long the process took and in what form is hard to determine. Initially, the interest in poultry was for eating (table fowl) and sport (cock-fighting).

The first table birds were apparently introduced by the Romans and were probably what are now known as the Dorking Fowl. These birds possess five toes and, over a long period, by careful selection, have become excellent table fowl. In *The Poultry Book* Wingfield and Johnson look upon the Dorking as *the English Fowl*. The breed forms the basis for many excellent crosses for table birds. Crossed with Indian Game they develop more meat without the objectionable yellow flesh of that breed.

The second principal breed but, in fact, the *oldest* is the Old English Game Fowl. Cock-fighting provided a major recreation from early times until made illegal in 1849. In fact, the sport continued to be practised until the early part of the twentieth century. There is still considerable interest in breeding and exhibiting Old English Game, the bantams being one of the most popular of all breeds.

Birds commenced to be imported from different parts of the World; from Asia, Italy, America, China and other lands. The growth of Poultry Shows, Royal patronage and the competition which was stimulated from trying to breed "perfect" birds, inevitably led to more poultry being kept. As a result "standards of perfection" were drafted and in 1864 the first Poultry Club was founded which proved "abortive", later in 1877 becoming the present Poultry Club.

The era of the poultry fancy had arrived and, despite cyclical fluctuations, has continued to grow in strength. Admittedly, there have been lean years when poultry keeping was regarded as being unworthy of attention. Yet such periods have been followed by revived interest in keeping birds for eggs or table, *or* just for fun!

The *Standards* stipulate recognised breeds and their colours. In Britain the Poultry Club gives recognition to any new breed and the *standard*; i.e. authorised description is published in *British Poultry Standards*. Since a new breed or variety cannot be introduced until strict rules are observed over a considerable period, it follows that only a serious breeder is likely to succeed. Even so poultry history is littered along the way with breeds which never survived after the death of the creator. Names like Black Marias, Bermuda Ruffles and Droodlies are examples of breeds which were created and then lost or revived under a different name. The first named appeared in *British Poultry Standards*, but was omitted when the breed became extinct and later reappeared as the Norfolk Grey.

* For a fuller account see *Modern Poultry Development*, H. Easom Smith, Spur Publications. This is believed to be the first History of Domestic Poultry Keeping.

ORIGIN OF PRESENT POULTRY

Although the *precise* origin of domestic poultry is unknown there is now general acknowledgement that the Red Jungle Fowl is the possible progenitor. Indeed, when looking at colours in existence it will be seen that the male Black Red (Black breast and tail, remainder Red) with his Partridge hen, is still a dominant breed or variety colour.

The multitude of colours known today came gradually, emerging from different parts of the world. In the hands of ardent fanciers, colours were standardised and improved upon. Selection and further selection, experiment following experiment; after many years the movement towards perfection began to gain ground.

Following Darwin and Mendel interest in poultry was stimulated in a variety of directions. At last there was a foundation upon which to build and to experiment. For commercial reasons much of the research has been directed towards the improvement of laying qualities or conversion to meat. Sadly this has resulted in the neglect of the standard breeds in favour of "hybrids"; i.e. breeding from close relatives and then crossing inbred lines.

Nonetheless, research carried out, notably by Professor Punnett* in Cambridge, has shown that certain "rules" may be applied:—

1. Whites may be of different types; e.g. may be blue bred, or may really be "constitutionally a cuckoo".
2. Barring is really a self-colour; e.g. black, to which a barring factor has been introduced.
3. Blue is not a true colour and this is why 100 per cent blue colour breeding is not possible. The true breeding forms are black and white splash, and black.
4. Blacks are generally straightforward and may be produced from breeding black with black.

However, in the experience of the author who has bred blacks for many years, there is a problem in getting blacks absolutely free from blemishes, such as white appearing in the wings. This is the result of blacks being blue-bred. If a purple-black factor is introduced the black may be more stable, but there are problems such as the wrong eye colour for Old English Game. A pale or yellow eye is not sound, red or dark being preferred.

5. Variations in feather colour such as spangling, lacing and barring are due to a special factor in each case.

Since then research has proceeded, especially in the U.S.A., but it is the fanciers, with little or no scientific knowledge, who have produced the multitude of colours now available. All that was required was a colour "bank" on which to draw, thus introducing different colours into the various breeds. Fortunately, with over 30 "colours", Old English Game gave most of the colour patterns available.

* See *Heredity in Poultry*, Punnett, R. C. MacMillan, London (out of print).

In the early soft-feather breeds such as Brahmas, Cochins, Andalusians and Leghorns many colours existed (and still exist). They also provided, along with the hard-feathered Game, the means for creating new breeds and colours. With judicious crossing, new breeds were created and new colours emerged.

COLOUR PATTERNS

There are many colour patterns found in poultry. These have been standardised over a period of some 100 years and, unfortunately, the pattern which has emerged can be confusing.

The *Black Red* is one of the basic colours (Black breasted, red male and partridge female), and this name is used for Game birds, but the same basic colour pattern is known as:

(a) Brown (Sussex and Leghorn)
(b) Partridge (Cochin, Plymouth Rock and Wyandotte).

The *Grey* is also referred to as a *Birchen* yet the two terms mean exactly the same thing.

The *Light* varieties in Brahma and Sussex are the same colour pattern as the *Columbian* in the Plymouth Rock and Wyandottes.

Blues are different in the various breeds, some being laced, others dark and yet still others quite light. *Lavender* may simply be a light blue or light grey. *Buff* may be a light fawn to a golden shade of considerable depth.

The description *Dark* may mean something different; for example, with Dark Dorkings and Dark Brahmas each is quite distinct in colour pattern. *Red* is a much laboured principal or subsidiary colour description yet this may mean a chocolate colour as in Rhode Island Red or a deep-red in the Black-Breasted Black Red of Old English Game. Often the term "mahogany" is used in the *Standards*, yet the colour is simply a vivid red as seen on the shoulders of Black Red Game.

Such are the problems of describing the many colours available and of understanding what the descriptions mean. The colour descriptions given to Old English Game can be very confusing and often defeat experienced fanciers, especially when dealing with "off colours". Space does not allow full explanation and readers wishing to study this subject more fully are referred to *Understanding Old English Game—Large and Bantams* written by the author.

Despite what is commonly thought—that colour was unimportant in Old English Game—much attention was paid to breeding "pure" colours, which was a sign of purity of blood. Admittedly, shape was of prime importance (and still is) but strains became famous by colour description: the Earl's Piles (Lord Derby's); Cheshire Piles, Parson's Gingers, and Yorkshire Greys were examples.

NOTES ON COLOUR BREEDING*

The beautiful range of colours and patterns within the individual feathers of various breeds of present-day poultry are one of the supreme examples of the way in which man can experiment to "improve" nature to comply with arbitrary standards. For in wild birds, and indeed in the jungle fowl from which our poultry is descended, nature is very economical with her patterns and where these are not of a haphazard nature to provide camouflage, they are only arranged on the part of the feather in full view. In poultry the fancier has managed to extend the pigments in many cases into the under-feather and has attained uniformity in colour and markings throughout not only the individual feather, but throughout the complete plumage as well.

It is perhaps this uniformity of colour and markings which make many breeds so strikingly beautiful and, certainly, in breeding them, the emergence of the new feathers is a source of constant interest.

Basic Colours

The colours found in the plumage of poultry are Black, White, Blue, Buff, Red and Yellow of various shades, and yet from this small list an enormous variety of patterns in various shades emerge.

Blacks and Blues

The self-Black bird usually has a metallic sheen of purple or beetle green—the latter being preferred in most breeds—giving a healthy alive look. If we add a factor to introduce white bars onto each feather we find that each feather can be barred with alternating bands of black with beetle green sheen and white of sky-blue tint as in the Barred Plymouth Rock.

Blue is really a dilution of Black in one of two ways. The first dilution factor is called Lavender and dilutes not only Black to Blue but also Red to Buff. The second dilution factor is an intermediary in the sense that one dose changes black to blue whilst the addition of a second dose dilutes the blue to a splashed white. As a result, the offspring of the mating of a splashed white and a black is all blue

Whites

Apart from the splashed white mentioned above true whites are also of two types—a dominant one and a recessive one. The former is caused by a factor which inhibits the formation of coloured pigment in its pure form but is far more successful in preventing the formation of black than it is of red. In fact, it has little success on red when only a single dose is present, but almost complete success on black. This is the factor which turns a black red into a pile and a Dark Indian Game into a Jubilee, and the reason why the breeder must go back to the black-red versions to preserve the red in the pile.

* Contributed by Dr. W. Clive Carefoot, a fancier who has made a special study of colour breeding.

The recessive white is an albino-type mutation in that such a bird lacks the means to produce pigment in the feather. Most heavy breed whites are of this nature and were derived as white sports from coloured birds.

Buffs

The exact genetic make up of Self Buff birds is not yet known. However, it is clear that several factors are involved. From the breeding methods it appears that there are dilutions involved, as self-blues and buffs have similar problems. It does not appear to be possible to deepen Buff into a Red without the emergence of Black into the feather and, in fact, Red is never found alone.

Black Reds and their Variations

The wild jungle fowl is a black red—and by this we mean essentially black everywhere except for red hackles, back, shoulder and wing bay. The red is of various shades and in modern breeds such as the Partridge Wyandotte Bantam the hackles are orange shading to lemon, the back is scarlet and the wing bay is bright bay. Coupled with brilliant beetle-green sheen the beauty of such a bird has to be seen to be appreciated.

We have seen that the addition of Dominant White to a Black Red makes a Pile; if, instead, we added a single dose of the Blue factor we would obtain a Blue Red.

Reds

The most popular breed of Red is the Rhode Island Red and this is really a Black-tailed Red, bred from the wild-type by the addition of Columbian factors. Specialisation for intensity of colour has deepened the Red into Mahogany in both the Rhode and the Golden Laced Wyandotte female, the latter being similar in make up to the Rhode but with the addition of factors adding a Black lacing to the edge of every feather, producing a fowl of outstanding beauty. In conjunction with the various factors previously mentioned one can alter the colours of the lacing to white or blue leaving the rich golden ground colour diluted only slightly to a rich deep buff in the first case, and unaltered in the second. We then obtain a buff with white lacing called Buff-laced and a "gold" laced with blue called illogically Blue-laced. These last three breeds can be bred together to form a pen of birds of really contrasting colours.

Variations in Colour

The golden Sebright bantam was evolved with lacing factors added to a Buff and this is the reason that Sebrights have laced tails and Wyandottes mainly Black.

The Black-tailed Red is an important basis on which to add various feather patterns. One can add black bars across the feather in various proportions, the Campine having three times as much black as gold and the

Golden Pencilled Hamburgh having equal widths. The Partridge Wyandotte has concentric rings of black on a "gold" ground, the outer colour being the ground colour. In this case the "gold" has been diluted to a soft partridge brown and a hen approaching the ideal, uniformly patterned from head to toe, and of even shade of colour throughout, is a real eye catcher.

Other combinations of pencilling and lacing on a Black-tailed Red Ground are present in the Indian Game and the Double-Laced Barnevelder, both of which have concentric rings around the feather starting at the edge of the feather with Black, the difference being that the Barnevelder has only two rings.

Unfortunately most of the patterns on a Black-tailed Red Ground are only of complete success on a female feather, with the males usually having a contrasting colouring of the Black Red type with some semblance of markings on the breast. For this reason Campines and Sebright Bantams have developed hen-feathered males.

A particularly widespread mutation is the one which changes Gold into Silver. This mutation has been of great importance in the commercial world for sex-linkage crosses to enable surplus males to be culled as hatched. The fancier has also made use of it in breeding the Silver counterparts of many golden breeds. For instance we have Silver Laced Wyandottes, Silver Pencilled Hamburghs, Silver Sebrights, Silver Campines and the equivalent of the Partridge Wyandotte called the Silver Pencilled Wyandotte. This factor also produces a Silver Duckwing from a Black Red and generally changes the Reds or Golds into Silver leaving the Black and its dilution Blue unaltered.

If one wanted to experiment there is no reason why one could not produce a "Silver Indian Game" with Black lacing on a silver ground, or a Wyandotte of Silver Ground edged with Blue instead of Black. The possibilities for a whole new range of colour combinations are virtually endless.

Apart from Barring, Pencilling and Lacing the two main pattern factors are Mottling and Spangling. The former is a recessive factor which adds a white tip to a black feather. The supreme example of this combination is of course the Ancona although other breeds such as Belgian Bantams have been very precise on occasions. Spangling is the addition of Black to the ends of Gold or Silver feathers, and is found on Spangled Hamburghs or "Moonies" as they were once known due to the circular shape of the spangles.

When the basic ground colour is gold of some shade, both spangling and mottling can be present at once giving a golden feather with a black bar near the end and a white tip. Unfortunately, the nomenclature of the breeds having these feather patterns is most confusing—in Sussex they are *Speckled*, in Old English Game, *Spangled*, and in Belgian Bantams they are known as *Millefleurs*. One of the most daintily coloured breeds is the Porcelaine Belgian Bantam in which the Millefleur has been diluted by the addition of the Lavender factor so that we have a ground colour of light straw and pea-shaped spots of pale blue tipped with white triangles. The overall effect has to be seen to be believed.

The silver edition of the Black-tailed Red is of course the Black-tailed White. Some of these Black-tailed Whites have varying amounts of Black in the neck and wings according to which Columbian restriction factors they carry. The most precisely marked of these is the Light Sussex, where a neck hackle of beetle-green-sheened black, surrounded by a clean clear-cut lacing of white, matches with complete balance a tail of intense black, with the same sheen.

At the opposite end of the scale is the Black-tailed White Japanese Bantam which has no visible black anywhere except the tail, and, in fact, should have white lacing on the tail coverts. They have, as indeed have all the Black-tailed White or Red breeds, black in the primaries and secondaries not visible unless the wings are opened.

Conclusion

There are countless variations of colour in our poultry and it is regretted that only the most precise, clear-cut patterns are able to be discussed here. Each breed has its enthusiasts, any one of whom will enlighten an interested enquirer as to the precise points of colour and markings required. The dedication of such fanciers will no doubt lead them to feel that their own particular fancy is the most beautiful bird in existence, and they will probably be correct for beauty is in the eye of the beholder!

GLOSSARY OF POULTRY TERMS

A.O.C.: Any other Colour.

A.O.V.: Any other Variety.

Atavism: Reverison to ancestral type.

Bantam: Miniature fowl, formerly accepted as one-fifth the weight of the large breed it represented; but nowadays about one-fourth.

Barring: Alternate bands of dark and light colour across a feather, as in Plymouth Rocks and Cuckoo varieties.

Bay: A reddish-brown colour.

Beard: A tuft of feathers on the throat under the beak, in such varieties as Belgians, Houdans and Polands.

Blade: The rear part of a single comb.

Booted: Feathered on shanks and toes.

Breast: In live birds, the front of the body from keelbone to base of neck. In dead birds, the flesh on the keelbone.

Cap: The back part of a fowl's skull. (Refers mainly to colour.)

Cape: The feathers at base of neck hackle, covering the shoulders.

Carriage: General deportment or bearing, especially when walking.

Chicks: Young birds recently hatched.

Chicken: In exhibitions, a bird of the current season's breeding.

Cinnamon: A dark reddish-buff colour.

Cobby: Short-coupled and round or compact in build.

Cock: A yearling male bird or older.

Cockerel: A male bird of the current year's breeding.

Cockerel-bred: Bred in line from matings specially designed to produce good exhibition cockerels. (Compare with pullet-bred.)

Coverts: The covering feathers on tail and wing.

Crest: A tuft of feathers on top of head.

Cushion: A mass of soft feathers covering root of tail, as in Cochins or Pekins.

Cushion Comb: An almost hemispherical comb, as in the Silkie. Sometimes called strawberry comb.

Daw-eyed: Having pearl-coloured eyes.

Diamond: The wing bay.

Double-comb: An expression usually meaning rose-comb.

Dubbing: Removal (by cutting) of comb, wattles and earlobes.

Duck-foot: Having the rear toe lying close to the foot instead of spread out, thus resembling the foot of a duck.

Earlobes: Folds of skin below the ears proper; sometimes called deaf ears.

Flat Shin: Flat fronts to shanks—a serious defect in most breeds.

Flights: Primary wing feathers, unseen when wing is closed.

Fluff: The lower or downy part of a feather; the soft fluffy feather on thighs of soft-plumaged breeds.

Foul Feathers: Mismarked or wrongly-coloured feathers.

Foxy: Rusty or reddish in colour.

Gay: Light or white in markings of plumage.

Gypsy-face: A face dark, purple or mulberry in colour.

Ground-colour: Main colour of body plumage, on which markings are applied.

Hackles: The narrow pointed feathers of neck and saddle, particularly in males.

Hangers: The lesser sickle feathers and tail coverts.

Hard feather: Close tight feathers as found on Game birds.

Hen-feathered: A male bird without sickles or pointed hackles (sometimes called a henny).

Hock: The joint between thigh and shank.

Horn comb: A comb with two V-shaped horns.

Keel: The breastbone, particularly its edge.

Knock-kneed: In at the knees.

Lacing: An edging round the margin of a feather—usually single: but when double the outer lacing is always round the margin.

Leader: The spike at rear of a rose comb.

Leaf comb: Shaped like a butterfly, or like two leaves, one each side of the head.

Lopped comb: A comb falling over to one side.

Mealy: Stippled with lighter colour.

Moons: Rounded tips or spangles to feathers.

Mossy: Blurred or indistinct.

Mottled: Marked with tips or spots of different colour.

Muff: Feathers on each side of the face (often accompanying a beard); sometimes known as whiskers.

Muffling: Beard and whiskers enclosing the face.

Mulberry-faced: Gypsy-faced.

Pea-comb: A small triple comb with three longitudinal ridges, as in the Brahma.

Pearl-eyed: Sometimes called daw-eyed; eyes pearl-coloured.

Pencilling: This has three forms: (*a*) concentric rings of fine markings following the outline of the feather (but not round its edge) as seen in Partridge Wyandottes; (*b*) narrow barring in Pencilled Hamburghs—the only breed in which this is called pencilling; and (*c*) the fine stippled markings found on females of Brown Leghorns and Partridge O.E. Game, etc.

Primaries: Main flight feathers.

Pullet: A female fowl of current season's breeding.

Pullet-bred: Line-bred from a mating calculated to produce good exhibition females. (See cockerel-bred.)

Reachy: Tall and of upright carriage and "lift", as in Modern Game.

Roach-back: Humped back.

Rose-comb: A broad comb with small spikes or "work" on top, finishing with a spike or leader at rear. Sometimes called a double-comb.

Rusty: Reddish-brown or foxy in colour.

Saddle: The rear part of the back, adjoining the tail. Covered by the cushion in the hen.

Scaly-leg: A defective coral-like condition of the leg, caused by an insect parasite.

Secondaries: The second group of main quill feathers on wing adjoining the primaries. Unlike the primaries, visible when wing is closed.

Self-colour: One uniform colour, unmixed with other colours or markings.

Serrations: "Saw-tooth" sections of a single comb.

Shafty: Lighter on stem of feather than the ground-colour.

Sheen: Bright surface gloss or black plumage. In other colours usually described as lustre.

Shoulder: The upper part of the wing nearest the neck-feather. Prominent in Game breeds, where it is often called the shoulder-butt.

Sickles: Long curved tail feathers on a male bird.

Side sprig: An extra spike growing out of the side of a single comb.

Single-comb: A single fleshy growth or blade on head.

Slipped wing: A wing in which the primary flight feathers hang below the secondaries when wing is folded; a condition often allied with split wing, in which the primaries and secondaries show a very distinct segregation in many breeds of bantams.

Smut: Dark or smutty markings where undesirable, such as in under-colour of R.I.R.

Spangling: Markings (usually round) on the tips of feathers.

Spike: The rear leader on a rose comb.

Spur: A horny growth on the inside of a bird's shank, pronounced in cocks and used as a weapon of offence.

Squirrel-tail: A tail projecting forward towards the neck.

Strawberry-comb: A lump comb resembling half a strawberry, as in the Malay.

Striping: The very important markings down the middle of hackle feathers, particularly in males of Partridge varieties.

Sword-feathered: Having sickles only slightly curved, or scimitar-shaped, as in Japanese.

Symmetry: Proportion; harmony in shape.

Tail-coverts: See coverts.

Thumb-mark: A faulty indentation on the side of a single comb.

Ticked: Plumage tipped with different colour; usually applied to V-markings as in the Ancona.

Tri-coloured: A fault in buff birds, the hackles, wings, body-colour and tail showing three different shades.

Trio: A male and two females.

Type: Mould or shape; see symmetry.

Under-colour: Colour of fluff beneath surface plumage, not seen until the feather is lifted.

Variety: A sub-division of an established breed. It is considered that type makes the breed, colour and markings the variety.

Vulture hocks: Long stiff quill feathers projecting backwards from the hocks.

Wattles: Fleshy appendages below the beak, strongly developed in most male birds.

Wing-bar: A band of dark colour or markings across the middle of wing.

Wing-bay: The triangular surface of secondary feathers as exposed when wing is folded.

Wing-bow: The part of wing between wing-bar and shoulder.

Wing-butt: The end of the wing; end of primaries.

Wing-coverts: Feathers covering roots of secondary feathers.

Work: The small spikes or working on top of a rose comb.

Wry-tail: A defective tail carried habitually to one side.

From *Bantams and Miniature Fowl* (3rd Edition) W. H. Silk, Spur Publications.

TECHNICAL TERMS
USED IN DESCRIBING FOWLS.

Black Minorcas in their heyday from a drawing by J. W. Ludlow (Other colours: White and Blue)

LARGE FOWL

Large Fowl vary in size, colour, and other characteristics to a marked degree with the result that more than 60 breeds have been standardised at one time or another. In the U.K. the governing body for poultry is The Poultry Club and the detailed descriptions of the breeds are contained in *British Poultry Standards* a book of approximately 400 pages.

These standard breeds vary in many ways and may be classified as follows:

(1) Light Breeds; e.g. Ancona, Leghorn, Old English Game, with weights of around 6 lb. or below for adult birds.

(2) Heavy Breeds; e.g. Brahma, Cochin, Dorking, Modern Game, Indian Game, with weights from just above 6 lb. to 12 lb. or more.

These weights are an approximate guide and inevitably there is some degree of overlap; e.g. some light breeds may exceed 6 lb.

Charles Francis

ANCESTORS OF MODERN POULTRY

There has been much speculation regarding the origin of our domestic poultry. Unfortunately the question has never been really answered—nor is it likely to be—for the simple reason that breeds of poultry have existed for many centuries. Even the period when poultry were first domesticated is lost in antiquity. Moreover, they have been found in very many parts of the world before adequate communication was established for transfer from one place to another.

Records exist of the keeping of poultry by the Egyptians more than 2,000 years ago. The Greeks and Romans also kept fowl and, indeed, it is recorded that Dorkings and Old English Game were in Britain at the time of Julius Caesar (c. 55 B.C.).

There are a number of Jungle Fowl all of which may have contributed to the development of domestic fowl. In the illustration (top) is shown the Red Jungle Fowl (*Gallus Bankiva* or *Gallus Ferruginous*). Many believe that this bird is the true progenitor, but others feel that the existing breeds vary so much in characteristics that it would be impossible for one small bird (around 3 lb.) to produce the Giants weighing 12 lb. or more.

The birds shown at the lower part of the illustration are the Grey Jungle Fowl (*Gallus Sonneratii*) and the Ceylon Jungle Fowl (*Gallus Lafayettii*). Other jungle fowl also exist, including *Gallus Giganteus* which may be the ancestor of the Malay.

Charles Francis

ANCONA

This breed came from Italy. A very active bird which is an excellent layer of white or cream eggs. Main colours are beetle green/black with white tips in profusion (V-shaped). A rosecomb species also exists.

ANDALUSIAN

Originating in Spain this bird is primarily a "layer". The main colour is slate blue with black lacing, which should be quite prominent.

ARAUCANA

Lavender

The breed standard requires blue-grey plumage as even as possible. Other colours are Black Red, Birchen, Brown Red, Golden Duckwing, Cuckoo, Pile, Black and White after the colouring of Modern Game. Originating in Chile this breed lays a blue or green egg.

ASEEL

Spangle

Male Aseels are reputed to be the original fighting cocks of India. The colour shown is a Spangle, but other colours exist such as Dark and Light Reds, Blacks, Whites, Duns (Blues), Yellows, Greys and Duckwings (see Old English Game).

AUSTRALORP

Originated in Britain as the Black Orpington, then was transferred to Australia. On its return it became the Australorp. There is only the single colour—Black with a brilliant green sheen.

BARNEVELDER

Double Laced

This Dutch breed lays brown eggs. The main colour is the Double Laced (illustration), but there have been Black, Partridge, and Silver colours. The lacing should be clear and even.

Charles Francis

BRAHMAS

Dark   Light

This is a very old breed which originated in India. There are three main colours: Dark, Light and White. They are magnificent birds, proud yet gentle. Note the feathered legs. Plumage is supposed to be closer than the Cochin. The comb is "pea" shaped; i.e. "triple", having three longitudinal ridges.

CAMPINES

Silver   Gold

The Campine is said to have been known in Belgium for around 500 years. A large white egg is laid. The two colours are as shown. Each body feather is barred with black which is approximately three times the width of the silver or gold.

COCHINS

Buff   Partridge

The older poultry books referred to the Cochin as the "cart horse" of the poultry fancy. It originated in China. The main colours are Black, Blue, Buff, Cuckoo, Partridge and White. Colours not shown are self-explanatory (for "Cuckoo", see Plymouth Rock). The *bantam* is thought to be a distinct breed and not a pygmy version of the large Cochin. There are many shades of Buff, but all should be free from blotches. A nice even rich buff is very attractive and should be the aim. The "Partridge" is the typical Black Red male and Partridge female. In some breeds this standard type is known as a *Black Red* (see Origins of Domestic Fowl).

Charles Francis

CROAD LANGSHAN

This breed has existed for more than 100 years in the U.K. It was imported by Major F. T. Croad. The colour should be a beetle-green black with no purple or other foreign tinge. There is also a white colour which should be free from any other colour. The breed lays large brown eggs and is a very commendable table bird. (See *Modern Langshan* which is a distinct breed.)

DORKING

Dark

The neck hackle may be white or straw coloured striped with black, in both sexes. In the male the black parts should be free from other colour (except sheen).

DORKING

Silver Grey

As implied by the description the white should be clear silver-white. In the *female* each body feather should be pencilled in a darker grey. The breast should be a red colour (some variation in shade is permissible).

Red

This is a beautiful colour. Ideally where "red" the redder the better, but, as shown, different shades exist.

Other colours are White and Cuckoo (*see* Plymouth Rock).

FAVEROLLE

Ermine

This colour is predominantly white with black in neck hackle and tail. In the hackle each feather has a black centre. A smutty colour on the back is regarded as a serious fault.

Salmon

This is probably the most popular colour. In the male the shoulders should be a "cherry mahogany", a colour which has been established in a positive fashion over the last 50 years (earlier illustrations showed buff shoulders). In the female the cream breast and thighs should be free from any other colour. The body colour is described as "wheaten brown" with darker stripes in the hackle.

Other colours available (many now rare) are Black, Blue, Buff and White. With all the self-colours clear, even colouring is required.

Charles Francis

HAMBURGH

Black

Rich, lustrous-black covered in a dark green sheen is the main requirement. (Note the rose comb which is a special point of beauty in this breed as well as the white ear lobes.)

Gold Pencilled

The male is an orange-red (described in the standards as "bright red bay") with broad tail feathers of black with a green sheen (encircled with gold). In the female the golden colour should be pencilled with black.

HAMBURGH

Silver Spangled

In both sexes the ground colour is silver-white with green-black "spangles". These vary in shape and size and on the body of the female are so frequent that they almost touch.

Special note: With the pencilled and spangled varieties "double-mating" is necessary—one pair to produce show pullets and the other to breed show cockerels. Other colours are Silver Pencilled and Gold Spangled.

HOUDAN

This crested and muffled bird is black with a green sheen which is mottled with white spots. The comb is termed "leaf" being shaped like two leaves in the form of the wings of a butterfly. The breed originated from France.

INDIAN GAME

Dark

The male bird is basically a green-black bird, hard-feathered and glossy. The wing-bay is chestnut and a *slight* amount of this colour may appear at the base of the hackle and across the back. However, winning birds tend to be very dark without the brown mixture in the hackle or on the back. In the female the body is dark chestnut brown with precise *double-lacing* in green-black. Deep-yellow legs and pea comb are essential.

Jubilee

Basically the black on the Dark Indians is replaced with white on the Jubilees. Double lacing on the female is still necessary, but often difficult to achieve. Mating Jubilee with the Dark colour is usually advocated.

31

Charles Francis

LEGHORNS

This Mediterranean breed was one of the finest layers. Sadly it lost ground, but is now returning.

Cuckoo

The plumage should be light blue or grey with darker bars. The latter are not absolutely distinct.

Buff

Some variation is allowed in the Buff colour. However, a sound even colour should be the aim.

Brown

The male bird is a pale Black Red and the female a very dark partridge with golden hackle striped with black. Sound breast colour is required in males and a good even brown in females. For this reason double-mating will be necessary to obtain show specimens of both sexes.

Pile

The predominant colours are white, orange and red as shown. They were once a popular colour and are worthy of revival.

Silver Duckwing

Golden Duckwing

The markings are practically the same. With the *Golden* colour the hackle should be light yellow or straw, the remainder of the back and shoulders being a golden colour. The Golden hen is darker than the Silver Duckwing hen.

Other colours are Black, Blue, White, Exchequer (black and white splash), Cuckoo, Mottled (black and white tips) and Partridge (as for Wyandottes).

Charles Francis

IXWORTH

In both sexes the plumage is white with the same colour legs (this is one major point where the breed differs from Indian Game which was used in its make-up).

JERSEY GIANT

Black

In the Black variety the plumage should be green-black with considerable sheen. The other colour available is White. The standard weight for a cock is 13 lb.—which is large—hence its name!

MARANS

Dark Cuckoo

As the description implies, the plumage is barred to resemble a cuckoo. The barring is blue black. This breed is famous for its dark brown eggs. Other colours are Black, Golden Cuckoo (golden and black bars), Silver Cuckoo (more white and lighter bars than Dark Cuckoo).

NAKED NECKS

Blacks
(Transylvanian Naked Neck)

Colours are numerous and include Blacks, Whites, Cuckoos and Reds. They have never been popular.

NEW HAMPSHIRE RED

The predominant colour is red with black appearing in the tail. American in origin they have never become as popular as the Rhode Island Red from which the breed was developed.

NORTH HOLLAND BLUE

This is a barred breed, the basic colouring being blue-grey with dark bars. The male bird is lighter in colour than the female. The barring is not as positive as in the other barred breeds, but the blue colour should be even and distinct.

Charles Francis

LA FLECHE

The plumage is green-black with lots of sheen. The horn-shaped comb is a characteristic of this French breed.

MALAY

Black Red

The male follows the normal Black Red pattern. With the female the body colour may be as shown (cinnamon) or a more definite partridge colour (*see* Old English Game).

MALINES

Cuckoo

The plumage is blue-white with bars of black. The shanks are lightly feathered. Other colours are Blue, Black, Gilded Black, Silvered Black, Gilded Cuckoo, Ermine (similar to Light Brahma) and White.

MODERN LANGSHAN

Black

Both sexes should be black with a green sheen. Note the tallness of the *Modern* and compare this with the Croad Langshan. Other colours are Blue and White.

MARSH DAISY

Wheaten or Golden

In the male the colour is Gold of different shades. There is some confusion regarding the breast which may be a golden brown or a stone colour (as shown). The female is rather like the old type of wheaten hen in colour—brown on top with breast a pale cream. A *Buff* also existed but seems to have become extinct. In old poultry books a White colour is also mentioned. Apparently the breed thrived on marshy soil—hence its name? However, the fact that the breed originated in the village of Marshside (Nr. Southport) may have some significance.

OLD ENGLISH PHEASANT FOWL

Gold

Many regard this breed as a variation of the Hamburgh and there is certainly a likeness. Variations in shades of colour seem to occur from a gold to a "rich bay". A *Silver* variety also exists.

Charles Francis

MODERN GAME

Black Red

The male has a black breast, orange hackle and red wing bows and back. The wing-ends (bays) should be a rich chestnut colour. Hens should be a good, even partridge with golden hackle striped with black (breast salmon). Eyes should be red and legs willow.

Pile

The cock is white with red as shown. In the female the body should be as white as possible with golden hackle and salmon breast (often chestnut). Eyes should be red and legs a deep yellow.

Birchen

The basic colours should be black and silvery white. Lacing should come half-way down the breast and be very delicate. Eye colour should be dark and the legs and beak should be black (or as near as possible). A dark (gypsy) face is desirable.

Brown Red

The colour formation is as for Birchens except the silver white is replaced by a rich bright lemon. In both Birchen and Brown Red the males are crow winged; i.e. black to end of wing.

Golden Duckwing

The male is black-breasted, with a cream-white hackle and golden/orange wing bows. The wing bays should be white. Females are a delicate steel grey with salmon breast and white hackle lightly striped with black. Eye colour is red and legs willow.

Silver Duckwing

This variety is similar to the Golden except the male has a silver white hackle and shoulders. The female is a lighter shade of grey on the body.

Modern Game are the birds kept by the dedicated fancier. Perfection in colour is essential with eye and leg colour also exact. Large Modern Game are now very rare; see *Understanding Modern Game*, Batty, J. and Bleazard, J. P., Spur Publications, where full descriptions may be found.

Charles Francis

OLD ENGLISH GAME CARLISLE OR SHOW TYPE

Black Red

Male: hackle and bays orange red, remainder (where indicated) bright red; breast, legs and tail black; wing bar, blue-black. Female: a good even partridge; golden hackle striped with black; breast salmon-brown (robin). When legs are white this colour is permissible in the tail of the Male bird. Legs may also be willow, grey or yellow.

Brown Red

Male: hackle yellow to orange; shoulder and back red-orange; crow winged. Female: black or black with brown markings and hackle laced (appears striped) with gold. Often the male has gold lacing on the breast and is strictly a *Streaky Breasted Orange Red*.

Greys

There are two colours, silver grey and black as shown. Otherwise they are as for Brown Reds.

Blues

Self blues should be a sound even blue all over. However, the cock usually has a darker colour hackle of dark blue/black or grey-blue. There are many variations which contain blue, e.g. Blue Reds. Blues are often described as "Duns".

Silver Duckwing

Male: silver hackle and shoulders; wing bays white; remainder black with wing bars a metallic blue. Female: body and hackle silver grey; hackle striped with black; breast salmon.

Golden Duckwing

As for the Silver, but Gold replaces the silver colour. The hen has a tinge of fawn in the silver grey body and tends to be darker. If too much "red" appears on the wing bows with the gold this is strictly a Birchen Duckwing.

Note: the Carlisle type of bird has been perfected in shape to comply with show standards. See *Understanding Old English Game,* Batty, J., Spur Publications.

41

Charles Francis

OLD ENGLISH GAME
OXFORD OR PIT GAME

Brown Breasted Brown Red

Male: ginger red hackle with red wing bows; breast mahogany or brown. Tail may be black (as shown) or a mixture of black and mahogany. Female: dark brown and black with gold lacing on hackle and breast (sometimes the breast is brown).

Black-Breasted Black Red or Dark Red

There are two colours dark red and black, often with a purple sheen. A gypsy face and black legs are desirable. The very dark birds are the true Black Breasted Black Reds. Often the female is practically black all over with a slight red tick in the hackle. The male is crow-winged.

Gingers

The male is a bright ginger red. Dark feathers appear in the wing and tail. Females may be a light "ginger" or a buff colour laced with bronze and black.

Pile

The bright orangey-red pile (shown) is known as a blood-wing pile. Custard and lemon piles also exist. The birds shown are "Tassels".

Blue Reds

(Also known as Red Duns) Similar to the Black Red shown on the preceding illustration, but blue replaces the black on the male. The female is really a blue-partridge colour.

Partridge Hennies

Cocks are "hen-feathered"; i.e. having no distinct sickle feather and having the same plumage as the hen. In the illustration the Partridge variety is shown. There are other colours such as Grouse, Black and Blue Reds.

Note: Cockfighting is illegal, but many fanciers have attempted to retain the old ideals and have avoided aiming for the perfect shape, believing that type would be lost. However, this thinking is not necessarily correct and the Carlisle type are often just as "gamey" in appearance.

Charles Francis

ORLOFF
Spangled

Male: hackles rich orange; wing bows mahogany; breast and tail black. The wing bay should be white. The female is light mahogany throughout. Both sexes should be evenly spangled. Muffling is white. Other colours are Black, Mahogany, and White. Originated in Russia.

ORPINGTON
Black

Plumage is black with a green sheen. This breed takes its name from the town of Orpington in Kent.

ORPINGTON
Blue

Colour is dark slate blue with a lighter shade on breast laced with the darker shade.

ORPINGTON
Buff

The buff colour should be as even as possible.

White is the remaining colour.

PLYMOUTH ROCK
Barred

Bars should be narrow of a beetle-green colour on a blue-white ground colour. Each feather should be barred straight across the same width as the ground colour space. The tip of each feather should be black.

Buff

The buff colour should be as golden as possible and evenly distributed.

The remaining colours are Black (beetle green), Columbian (as Light Sussex shown later) and White.

Charles Francis

POLAND (POLISH)

Gold Variety

Plumage is golden bay with black markings. Note the pronounced lacing and large crest. Male and female are muffled and the horn-type comb is very small.

Silver Variety

Description as for Gold, but silver the main ground colour.

White Crested Black

There are two colours—metallic black and white. No muffling appears on this variety. Black must appear on the front of the crest (as illustrated).

White Crested Blue

As for White Crested Black, but blue instead of black.

Other colours in Poland are Chamois (buff with white lacing) and White (a self colour).

REDCAP

(Derbyshire Redcap)

The male is a Black Red with fringes and tips in black on the red feathers. The female has a half-moon spangle on each feather. Some variation in colour exists and there is a tendency to look for a darker coloured plumage (approaching claret).

RHODE ISLAND RED

A very deep red is required (almost chocolate coloured). This should be as even as possible with a brilliant gloss overall. There is also a rose-combed variety.

Charles Francis

SCOTS GREY

The ground colour is steel grey with each feather having black bars running across, cuckoo-like. In the female markings tend to be heavier.

SCOTS DUMPY

Cuckoo

This breed has the familiar cuckoo markings. Other varieties are Black, Dark and Silver Grey (for the last two see the colour of Dorkings). This is a very rare breed. Despite its short legs the male reaches a weight of around 7 lb.

SILKIES

Blue   *Buff*

This rather small fowl (around $2\frac{1}{2}$ lb.) is very popular as a sitter, being persistently broody. The plumage is fluffy, almost like fine hair. *Blues* are an even shade of blue-grey, and Buffs a golden-buff colour. With the profusion of feathers down to the toes these birds are rather quaint. Despite their smallness they are classified as large fowl.

Other colours are Black and White.

SPANISH

The colour of the plumage is black with a dark green sheen. A characteristic of the breed is the large white face which must be kept in top condition for showing.

SUMATRA GAME

The plumage is green-black with the maximum of sheen. The face is dark (gypsy faced) and the comb the pea-type (any other type is a serious defect).

Charles Francis

SUSSEX

Light Sussex

One of the most popular of varieties the Light Sussex is predominantly white with hackles striped with black. For show birds the hackle is required to be very "bold". The tail is black.

Speckled

The main colour is rich dark mahogany with green-black stripes and white tips. Tail feathers are black with white tips.

Buff

A good even colour is required of golden buff. The neck hackle should be striped with green-black. Tail is black.

Red

A deep red is the main colour. Tail is black and neck hackle is striped with black.

Brown

This is another variation of the Black Red, but with a rich deep mahogany for the male's neck hackle. The female is brown with the lighter hackle striped with black. Breast is a lighter brown. Tail in both sexes is black.

Silver

This colour is rather like the Grey in Old English Game. It differs in the male having silver wing bays and the female having more lacing on back and breast.

A self-White also exists.

Charles Francis

SICILIAN BUTTERCUP

Golden

In the male the predominant colour is bright orange-red running into a deeper colour below (red bay). Spangles may also appear on the base of the neck (not shown in illustration). The female is golden buff covered in black spangles, except on neck. Both sexes have black tails with gold or bay.

Other colours are Brown (a Black Red with black stripes in hackle and female golden-yellow striped with black but breast salmon), Silver (as for Golden but silver colour), Golden Duckwing and White.

WELSUMMER

With slight variations both male and female follow a typical Black Red pattern. The female is darker than the typical partridge and her breast is rich chestnut red. Markings appear on the male and female as shown.

YOKOHAMAS

Red Saddled

The main colours are red and white with white lacing (or spangling) on the breast and on the saddle of the hen. The red may also run onto the thighs. A walnut comb is necessary. A White colour is also standardised.

SULTAN

The plumage is pure white throughout. A characteristic feature is the vulture hock, i.e. feathers growing backwards from joint between thigh and shank. Both a crest and muffling should be present.

PHOENIX

Black Red

This is a variety of the Japanese long-tailed fowl although in this country they never seem to achieve the very long tails. According to Mr. M. A. Beaumont from Tasmania the true long-tailed fowl is the Onagadori. He saw this breed in Japan with tails of 30 feet long (see *Poultry Club Year Book,* 1967). The plumage description is as described for Old English Game—Black Red/Partridge.

PHOENIX

Golden Duckwing

The colour description is the same as for Old English Game of this colour. Other colours include Spangles and Silver Duckwing as for Old English Game.

Note: in the earlier *Poultry Standards,* Phoenix and Yokohama were grouped together under "Yokohamas". Now the single combed birds are segregated under "Phoenix".

Charles Francis

WYANDOTTES

Gold Laced

As shown, the colour is really a golden bay colour in varying shades with hackle striped with black. The body is single laced with black. There are differences in the shades achieved but this is not critical. The tail is a greeny-black.

Silver Laced

As for Gold Laced, but ground colour silver white quite free from any discolouration.

Columbian

The main colour is white with distinct black stripes on the hackle, clearly circled with white.

Partridge

Although a variation of the Black Red the male tends to be rather yellow on the hackle with a black stripe. The female is finely and clearly pencilled in black on each feather; a very important requirement.

Blue Laced

This is a beautiful colour being similar to the Gold and Silver Laced, but with blue lacing. Contrary-wise the fancier calls this Blue Laced but it is really a Gold with Blue Lacing.

Silver Pencilled

This is similar to the Partridge except a silver colour replaces the yellow and red and in the female the grey appears instead of the partridge brown.

Other colours which exist (or have existed) are White (very popular), Black, Blue, Buff, Buff Laced and Red.

Charles Francis

SOME RARER BREEDS

BRESSE
Black

This French breed is available as Black (shown) and White. Blues, Buffs, and Greys have also been bred, but are not included in the British Standards.

CREVE-COEUR

Again a French breed this is one of the oldest in Europe. It has a crest and the comb is of the horn type. Black is the colour standardised.

LAKENFELDER

There is only one variety of this fowl which at various times has been referred to as Dutch, German and even Belgian. The hackle is a velvety black, extremely glossy, and the remainder of the body is pure white. These are very handsome birds.

NORFOLK GREY

The colour follows very closely the Grey of Old English Game. In both sexes the base colour is black with the cock having silver-white hackles, shoulders and wing bows. The female's hackle is silver-white. In both sexes there are black stripes in the neck hackle.

GUINEA FOWL

Guinea Fowl originate from West Central Africa although, in fact, they are quite wide-spread in Africa and nearby islands. In recent years their commercial possibilities have been realised and there is a movement towards keeping them on a small- and large-scale basis. The *Pearl* is a dark grey with small spangles ("pearled"). The *Lavender* is a soft, light grey also spangled. Other colours exist such as the White and "Splashed".

GAME FOWL IN NATURAL SURROUNDINGS

The large Old English Game shown are of the Oxford type. They are "Light Reds" or, more accurately, Black Red, Partridge Bred, with white legs. A famous strain was the Derby Reds, bred by the then Earl of Derby up to his death in 1835. It is recorded that this strain was bred at Knowsley for more than 100 years.

In their natural surroundings all types of poultry look healthy, colourful and full of vigour. They can scratch and forage for food which keeps them fit. In such an environment their beauty is unsurpassed, rivalling even the most outstanding of wild birds.

THE ENGLISH GAME COCK
by Herbert Atkinson

Small head, and strong and lofty neck,
Hooked beak, and bold large eye;
His breast, and back both broad and flat,
Short round and lusty thigh;
With strong clean shanks, and tapering toes,
And strong tail carried high.
Wings that are powerful, large and long,
Thin sharp spurs, set on low;
And lofty mien that indicates
Desire to meet the foe.

In hand so hard, and strong, yet light
Balanced in every part,
Belly, and fluff he's next to none,
Yet amply plumaged too,
That glows and glistens in the sun,
With many a beauteous hue;
While every action shows a grace, agility, and pride,
And courage that will last as long as flows life's ebbing tide,
As it has shown in countless sires of ancestors beside.

From: *Cockfighting and Game Fowl.* Herbert Atkinson.

TYPES OF FEATHERS.

1. Golden Pencilled (Hamburgh Sickle). 2 and 3. Golden Pencilled (Hamburgh Hen Feathers). 4. Partridge (Cochin). 5. Silver Laced (Wyandotte Cock's Hackle). 6. Silver Laced (Wyandotte's Hen's Breast). 7. Silkie. 8. Double Laced (Indian Game Hen). 9. Spotted (Guinea Fowl). 10. Barred (Plymouth Rock Hen). 11. Barred (Plymouth Rock Cock's Hackle). 12. White Spangle (Houdan Hen). 13. White Spangle (Houdan Cock's Hackle). 14. Black Spangle (Silver Spangled Hamburgh Sickle). 15. Black Spangle (Silver Spangled Hen's Breast). 16. Black Spangle (Silver Spangled Cock's Hackle).

BANTAM SECTION

Bantams are small fowl which are approximately one-quarter the size of the equivalent large breed, where this exists. Accordingly, weights are around 40 oz for the heavy breeds such as Indian Game and Sussex and in the region of 22 oz for breeds like the Ancona and Scots Grey. Many fall within this range of 22–40 oz; breeds like the Orpington, Rhode Island Red and Marans come around 30 oz.

Many of the bantams have equivalent breeds in large fowl. Accordingly, colour-wise there is very little difference so a guide to the Bantams can be obtained from consulting the large fowl colours. For this reason they are not repeated in the Bantam section.

Natural bantams having no large counterpart have been included. Some, like Old English Game, now differ so much from the large fowl that they have also been included. Cochins are regarded as a separate species even though large Cochins of the same colours do appear in the *Standards*.

Modern Game deserve special mention. They are very popular and have continued to develop long after the large Modern have gone out of fashion. Accordingly, they could justifiably be included in this section. However, the colours are exactly those shown for large Moderns. In addition, they are already portrayed in *Understanding Modern Game,* Batty, J. and Bleazard, J. P., Spur Publications. Readers requiring details on shape and size are referred to that source.

Frizzles are seen more often in Bantams than Large and, therefore, are included in this section.

Charles Francis

BOOTED

This breed is believed to be of very ancient origin. Presumably the name came from the heavily feathered shanks. In the White (illustrated) the plumage should be brilliant-white. Some strains are muffled. Other colours which existed were Black, Black Mottle, Millefleur and Porcelaine (*see* Belgian Bantams, overleaf).

COCHIN OR PEKIN

The birds illustrated are the Black variety. Other colours are as listed for large Cochins. In addition, there is the Mottled which is black with white at the tip of each feather.

FRIZZLES

A self-Blue pair of Frizzles is shown. The characteristic of this breed is that each feather curls back. Many colours exist: Whites, Blacks, and Buffs as well as many of the colours given for Old English Game. These are not natural bantams.

JAPANESE

Greys

The colour is similar to the Birchen Grey in Modern Game described under Large Fowl.

JAPANESE

Black-Tailed White

These are white with black tails and, as shown, the tail feathers are laced with white.

JAPANESE

Cuckoo

The ground colour is light grey with darker bars forming the cuckoo colour.

Other colours are Black-tailed Buffs, Buff Columbian (*see* large Wyandotte for general description of Columbian), Mottled, Red, Tri-coloured, Black Red, Brown Red, Golden Duckwing and Silver Duckwing.

Charles Francis

BELGIAN BEARDED BANTAMS

Barbu d'Uccle
Millefleur

Orange red dominates with mahogany red on the wing bows in the male. As shown, in both sexes there is an intricate pattern of black spots and white triangles. The tail is black.

Barbu d'Uccle
Porcelaine

The ground colour is described as "light straw" and the spots are blue with white triangles.

Barbu d'Uccle
Self Blue

This is a true breeding pale blue of an even shade throughout. "Grey Blue" is an appropriate description.

Barbu d'Anvers
Cuckoo

The plumage colour is light grey with bars of a much darker shade.

Barbu d'Anvers
Quail

This colour is a mixture of black, pale buff (umber) and gold. The black feathers in tail and back have gold shafts.

Barbu d'Anvers
Black Mottled

The main colour is green-black with an even distribution of white tips on the feathers.

Because of the complex descriptions of these colours, readers are advised to study the *Poultry Standards* very carefully. A very brief outline is given above. The Barbu d'Uccle is single-combed and feather legged whereas the Barbu d'Anvers is rose-combed and clean legged.

Many other colours are standardised such as Blue Quail, Black, White, Laced Blue, Blue Mottled, Ermines, Fawn Ermines, Partridges, Silvers, Spangles and Golds.

Charles Francis

OLD ENGLISH GAME BANTAMS

Black Red/Partridge

The Black Red should follow the colour of the Large Game. The breast of the cock is black and the hackles orange red. The shoulders are a dark, bright red and the wing bays are chestnut. The hen is partridge with a salmon breast and a golden hackle (striped).

Black Red/Wheaten

This is a lighter shade than the Black Red/Partridge for the Male. whereas the female is a creamy white, with pale orange hackle (this may be lighter than shown).

Spangles

These are similar to the Black Red/Partridge variety except:
(a) cock tends to be a darker red (burgundy);
(b) hen is a dark partridge;
(c) both sexes have an even distribution of spangles.

Pile

The male has hackle and shoulders a bright red with wing bars white and bays chestnut; the remainder white. The female is white or cream with salmon breast and striped or laced hackle (yellow lacing).

Golden Duckwing

Silver Duckwing

Duckwings are as described for Large Game. They take their name from the blue metallic bar on the wing of the male. The hackle of the male of the Golden Duckwing may be "off white" or a positive yellow colour. Yellow-straw colour should be the aim.

Although Bantam *Standards* are supposed to be similar to the Large Game, in practice the two differ quite considerably. Bantams are "cobby" birds with rounded bodies and short tails, often with very small sickle feathers. See the large Old English Game colours for other varieties.

Charles Francis

NATURAL BANTAMS
NANKINS   RUMPLESS

Golden Duckwing

The predominant colour is yellowy buff with the male having a darker tail of bronze colour.

Also known as "Rumpless Game Bantams". The colours follow the same pattern as for Old English Game.

ROSECOMB   ## ROSECOMB

Black   *Blue*

The self-black colour is covered in a brilliant green sheen. In this breed large, perfect white ear lobes are necessary. Black and White Rosecombs take the place of Hamburghs in bantams.

A sound medium shade of blue is required although in the male a darker shade occurs on hackle, back and shoulders.

A White variety also exists.

SEBRIGHT

Silver   *Gold*

Ground colour is silver white and around each feather is precise, black lacing.

Ground colour is gold and around each feather is precise, black lacing.

Study of the illustrations will show that the male is "hen feathered"; i.e. the same colour as the hen with no distinct sickle feathers. Although the original breed had purple face, wattles and comb, this is now difficult to achieve so it appears that bright red will become the standard.

TYPES OF COMBS

1. STRAWBERRY	5. SINGLE (UPRIGHT)	8. SINGLE (DROOPING)
2. ROSE	6. REDCAP	9. V
3. PEA	7. LEAF	10. HORN
4. HELMET		

DUCKS, GEESE AND TURKEYS

Ducks originate from the Mallard or wild duck, with the exception of the Muscovy which has distinct origins believed to be South American. The duck nearest to the Mallard is the Rouen.

All breeds have distinct characteristics of their own, but for table use Aylesbury or Pekin (or a cross of these) is recommended. The Rouen is also a large bird suitable for fattening. The major layer is the Khaki Campbell, but the Indian Runner is also a good layer.

The different breeds of **geese** have been developed from different species of wild geese. They are grazers and, accordingly, vegetarian in character. Many live to a ripe old age (40 years). For people with adequate grazing geese can be quite profitable—they also make excellent "watch-dogs".

Turkeys are traditionally associated with Christmas, although a modern development is the creation of smaller birds (mini-turkeys) which have a demand at other times of the year.

Charles Francis

DUCKS

AYLESBURY

This overall white bird is noted for its fine qualities for table use. Standard weight is around 10 lb.

BLACK EAST INDIAN

This is a very small duck and is more ornamental than "utility". The colour is green-black.

CAYUGA

This is similar to the Black East Indian duck except it is about 4 times as large (around 8 lb).

CRESTED

The characteristic feature of this duck is the crest. A pure white is shown, but there are other colours.

DECOY

Brown

The description "Brown" is something of a misnomer because the colours follow closely those of the Mallard. There is also a White variety.

INDIAN RUNNER

Fawn and White

The two colours are as indicated with the drake having the deeper fawn. Other colours which exist are Black (as for Black East Indian), Chocolate, Fawn (self-colour shade as for Fawn and White) and White.

Charles Francis

KHAKI CAMPBELL

The dark parts are a bronze colour whereas the remainder is khaki. Legs and webs are bright orange in the drake and a more subdued colour in the duck.

In "Campbells" there is a Dark variety and a White.

MAGPIE

As implied by the name this variety is a mixture of black and white in the fashion of a Magpie.

MUSCOVY

Black and White

These come in variations of Black, White and Blue, either as a self-colour (e.g. Blue only) or as shown in the illustration, a mixture of Black and White.

ORPINGTON

This breed was created by W. Cook who also produced the original Orpington fowl. The colour should be even buff throughout. The standard stipulates a "rich" shade, but variations exist and the colour in the illustration is probably typical.

PEKIN

The predominant colour is creamy-white. At one time a yellowy colour was preferred, but this has given way to the lighter shade.

ROUEN

These are very beautiful ducks and are coloured just like the wild Mallard. The drake has a dark green head with a white wing separating a deep claret breast. The rump and back are rich green black whereas the wings and remainder are French grey with fine lines. The duck is brown with darker pencilling on each feather. In both sexes the wild-duck wing bar of steel blue is present.

75

Charles Francis

GEESE

BRECON BUFF

In colour the aim is to obtain a buff which is a good solid coverage, although the gander is paler than the goose.

CHINESE
Brown or Grey

Chinese geese are majestic creatures which combine the colours of brown, fawn, grey and white as shown in the illustration. A White variety also exists.

EMBDEN

This is a pure white goose which has a long, broad body and a long neck. It does not have the prominent keel of the Toulouse, although around the same size: goose 20 lb—gander 30 lb.

ROMAN

The Roman is another White goose, but is smaller than the Embden. It weighs around 12 lb.

SEBASTOPOL

This is a "frizzled" goose having its white feathers curled in a distinctive way.

TOULOUSE

This is a massive goose. It has dark, steel-grey plumage with the feathers laced with white. The pouch shown extending below the bill is known as the "gullet". Although the standard calls for *steel-grey* many Toulouse now have a brown tinge to the feathers, softening the tone of the grey (may be due to a cross with another breed).

Charles Francis

TURKEYS

BRONZE

The plumage is a metallic bronze with markings as shown.

NORFOLK BLACK

The breed is a rich glossy black.

BLUE

The colour should be ashy blue of an even shade, free from lacing.

BUFF

The plumage is fawn or pale buff; in the Standards it is called "cinnamon brown".

79